NISTIR 7614

THE ABC's OF STANDARDS ACTIVITIES

This report supersedes NBSIR 87-3576

August 2009

Maureen A. Breitenberg
Standards Services Division
Technology Services
National Institute of Standards and Technology
Gaithersburg, MD 20899

Foreword

The Standards Services Division (SSD) within the National Institute of Standards and Technology (NIST) publishes information related to standards and conformity assessment as a service to producers and users of such systems—both in the government and in the private sector. This report provides a basic introduction to the U.S. standards system; explains what is meant by the term, "documentary standard;" and provides an overview of the U.S., international and regional standards systems. In addition, it describes the principles used in effective standards development efforts.

This report discusses the role played by private sector entities in the standards process, including the role played by the American National Standards Institute (ANSI). It also discusses the U.S. Standards Strategy (USSS), which provides a roadmap for reliable, market-driven standards in all sectors.

In addition, it covers some obligations of U.S. federal agencies with respect to their use and adoption of standards. In particular, this document covers the impact of the National Technology and Transfer and Advancement Act (NTTAA), the Office of Management and Budget (OMB) Circular A-119, *Federal Participation in the Development and Use of Voluntary Consensus Standards and in Conformity Assessment Activities*, as well as the obligations imposed by international and regional trade agreements.

The report is intended to provide basic information to help educate U.S. government agency officials, legislative staff, industry, and other standards professionals to make informed decisions regarding standards development policies and the effective allocation of limited standards development resources. It is also hoped that this report will serve as a starting point for further study and discussion on some of these important issues. Readers may also be interested in other related publications which are available on the NIST website at:
http://ts.nist.gov/standards/information/osc.cfm

Table of Contents

Introduction 1

Historical Notes on Standardization 3

Types of Standards 4

Standards Development Principles 8

The U.S. Standards System 10

U.S. Private Standards Developing Organizations (SDOs) 11

Role of the American National Standards Institute (ANSI) 13

Role of NIST ....

The International Standards System 14

Regional Standards Systems 16

Standards Usage/Conformity Assessment 21

Benefits and Problems of Standardization 22

National Obligations with Respect to Standards 24

International Obligations with Respect to Standards 25

Challenges Facing the U.S. Standards System 26

U.S. Standards Strategy (USSS) 29

Summary 30

Appendix I – Acronym Index 31

Acknowledgments

I would like to thank Anne Meininger, Mary Saunders, Ileana Martinez, JoAnne Overman, Carmina Londono, and Brenda Umberger for their careful review of and comments on this document.

                Maureen A. Breitenberg
                Standards Coordination and Conformity Group
                Standards Services Division

Abstract

This report is an introduction to standardization and the U.S. standards system for readers that are not familiar with this topic. It highlights some of the more important aspects of this field; furnishes the reader with both historical and current information; describes the importance and impact of standards development and use; and serves as guidance for using available documents and services. This report discusses some of the complexities of the U.S. standards system. It also provides some understanding of what is meant by a documentary standard, as well as an overview of the international and regional standards systems. In addition, it describes the U.S. Standards Strategy and the principles used in effective standards development efforts.

This report also discusses the role played by private sector entities in the process, including the role played by the American National Standards Institute (ANSI). It covers some obligations of U.S. federal agencies with respect to the use and adoption of standards. In particular, this document discusses the impact of the National Technology and Transfer and Advancement Act (NTTAA), the Office of Management and Budget (OMB) Circular A-119, *Federal Participation in the Development and Use of Voluntary Consensus Standards and in Conformity Assessment Activities,* as well as the obligations imposed by regional and international trade agreements.

INTRODUCTION

Standards have become an increasingly powerful force in the protection of public health, safety, and the environment; the development and commercialization of new technologies; and the facilitation of national and international commerce. Standards are so universally used that they are often taken for granted. While some might consider them to be about as interesting as watching grass grow, without them, modern life as we know it would unravel.

For example, consumers expect that 35mm camera film marked with the film speed designation, ISO 100, will likely give good photographic results if the camera's film speed is set at 100. However, few consumers understand that this is only possible because the film conforms to a standard established by the International Organization for Standardization (ISO), an international standards developing organization headquartered in Geneva, Switzerland.

While driving, we look for red, hexagonal stop signs; not blue, square-shaped stop signs. We know that inverted triangles indicate where traffic should yield. Yet few have heard about the Manual on Uniform Traffic Control Devices (MUTCD) or the State Highway Signs Book published by the Federal Highway Administration (FHWA) within the U.S. Department of Transportation. These documents contain national standards and guidance for traffic control devices and highway signs.

When buying home insulation, most consumers check the product's R-value. Few realize that the Federal Trade Commission (FTC) has established standards to ensure that such information is available to consumers and that it is based on standardized test procedures.

These are just a few examples of how standards affect everyday life. Standards are not only important for consumers, they are also critical underpinnings for business. In today's competitive world economy, ignoring the importance of standards can be a costly strategy for industry and government. Companies and entire industries may become less efficient. Transactions may become more costly in both dollars and resources necessary for buyer-seller negotiations. Markets can fragment as divergent requirements for products and services are developed and imposed.

Standards promote efficiency in domestic and international markets. By adhering to agreed upon standards, businesses can use widely accepted requirements and specifications to negotiate deals for products or services, avoiding contract ambiguities that might otherwise undermine such transactions.

Standards promote understanding between buyer and seller and facilitate mutually beneficial commercial transactions. Product[1] attributes cannot always be evaluated by individual purchasers by merely looking at the product or even from prior experience. There may be simpler communities where all products are made locally and where everyone knows the quality and performance characteristics of the products made by each and every producer. However, in most marketplaces, buyers are unable to make competent judgments regarding product characteristics and performance

---

[1] The term "product" is used in this paper to refer to a product, process, service, system, or personnel qualifications.

without assistance. Most products have become far too complex; and, in our global marketplace, suppliers are often unknown entities.

A product's conformance to accepted standards readily provides an efficient method of conveying complex information on the product's suitability. Architects can use standards in a shorthand manner when drafting plans for buildings. Procurement agents can also use standards as an easy way of communicating their needs to potential suppliers. Standards are used to replace large quantities of complex information needed to facilitate marketplace transactions. Hence, standards have tremendous economic impact on companies, nations, and even on the economic fabric of the world market.

Standards also underlie all mass production methods and processes. They promote more effective and organized social interaction and are essential in efforts to improve product safety and provide a cleaner environment. Standardized parts can reduce inventory requirements, facilitate product repairs, and allow interoperability between different products and systems. Standards can also promote competition by facilitating the comparison of prices of standardized commodities. In addition, standards can facilitate the introduction of innovative products and new technologies.

Today, an estimated 80 percent of world merchandise trade is affected by standards or regulations that reference or incorporate standards.[2] Standards are fundamental to the U.S. economy and vital to world commerce. In fact, the American Society of Mechanical Engineers (ASME) ranked the promulgation of standards among the top ten engineering accomplishments of the last century. Standards shared top-ten honors with such accomplishments as the inventions of the automobile and airplane.[3]

Standards permit society to make more effective use of limited resources and allow improved communication among all parties involved in particular activities, transactions, or processes. Indeed, standards are crucial to every form of scientific and industrial process. However, when standards are poorly written, they can cause significant economic damage. Poorly written standards can raise transaction costs, reduce product safety and quality, and create barriers to trade. They can also constrain innovation; entrench inferior technologies; and impede the development of interoperable products and systems.

Because standards have such an impact, it is imperative for decision makers and others to have some familiarity with standards and how they are developed. It is also important for decision makers to understand some of the significant standards-related issues that are faced by U.S. industry and the U.S. standards system.

---

[2] Dr. Arden L. Bement Jr., Director, National Institute of Standards and Technology, Forum on New Directions in Manufacturing, National Academy of Sciences, March 27, 2003.
[3] Ibid.

HISTORICAL NOTES ON STANDARDIZATION

The history of standardization is both fascinating and demonstrative of the scope and variety of standardization activities. One of the earliest examples of standardization is the creation of a calendar. Today we take the precision of our calendars for granted, unaware of the long history that touches not only Western civilization, but also extends back thousands of years to China, India, the Egypt of the Pharaohs, Arabia, and Mesopotamia.[4]

A predecessor of the American National Standards Institute (ANSI) noted that another of the first known attempts at standardization in the Western world occurred in 1120. King Henry I of England ordered that the ell, the ancient yard, should be the exact length of his forearm, and that it should be used as the standard unit of length in his kingdom.[5]

That history also notes that, in 1689, the Boston city fathers recognized the need for standardization when they passed a law making it a civic crime to manufacture bricks in any size other than 9x4x4 inches. The city had just been destroyed by fire, and the city fathers decided that standards would assure rebuilding in the most economic and fastest way possible.[6]

Eli Whitney is sometimes referred to as "the Father of Standardization" in the area of interchangeability, having originated and implemented the concept of mass production in the United States in 1780. He was awarded a contract to produce 10,000 muskets by then Vice-President Thomas Jefferson. Though standardized parts had been successfully used in other parts of the world, Whitney brought the concept to the United States when he divided the manufacturing process into individual steps and put different groups to work on each step of the process. All parts of the same type were copied from a model musket and were made to be interchangeable. Subsequently, when he appeared before the U.S. Congress with a collection of assorted parts and proceeded to assemble ten working muskets by selecting the required parts at random, the U.S. Congress was convinced of the benefits of mass production made possible by standardization.[7]

Standards are known to have existed as early as 7000 B.C. when cylindrical stones were used as units of weight in Egypt. However, it was the great blaze in downtown Baltimore in February 1904 and other similar catastrophes that provided tragic and undeniable evidence of the importance of standards within the United States. The fire razed a large section of Baltimore for more than thirty hours, destroying 1526 buildings covering more than seventy city blocks. All electric light, telephone, telegraph, and power facilities were also razed. Fire companies came from as far away as New York to battle the blaze, but many of the reinforcements were useless because their hose couplings did not fit the Baltimore hydrants.[8] In contrast, 23 years later, help from 20 neighboring

---

[4] National Institute of Standards and Technology, "A Walk Through Time," http://physics.nist.gov/GenInt/Time/time.html.
[5] American Standards Association, "Through History with Standards" in Rowen Glie (ed.), Speaking of Standards, Cahner Books, Boston, MA, 1972, p. 38.
[6] Ibid., p. 42.
[7] Ibid., p. 44.
[8] Rexmond C. Cochrane, Measures for Progress: A History of the National Bureau of Standards,

towns saved Fall River, Massachusetts from destruction since hydrants and hose couplings had been standardized in those communities.[9] Today, standards for compatibility and interoperability have resolved a myriad of similar challenges.

As late as 1927, color-blind motorists had as good (or as bad) a chance as anyone else when trying to interpret traffic signals. Purple, orange, green, blue, yellow, and red lights greeted motorists as they drove from state to state. In some states, green meant "Go;" in others "Stop." Red (not yellow) lights meant caution in New York City. In 1927, a national code for colors was established through the work of the American Association of State Highway Officials (AASHTO), the National Bureau of Standards (now the National Institute of Standards and Technology or NIST) and the National Safety Council.[10] Imagine the chaos that would occur now during any major U.S. city's rush hour if newcomers and tourists did not know what traffic signals meant!

Probably the most significant standard ever developed in the United States was the railroads' standard track gauge. By 1886, it had become the U.S. standard. Legend traces the origin of the standard gauge back to the coalfields of northern England and its evidence of rutted roads marked by chariot wheels dating back to the Roman Empire. This legend may have some validity since the wheels of horse-drawn vehicles were often approximately 5 ft (1500 mm) apart to accommodate a carthorse in between the shafts.[11] However, whatever its origin, today this standard enables railroad rolling stock to cross the country.[12]

However, it was World War II that created an urgent need to harmonize standards at the international level. Allied supplies and facilities were severely strained because of the incompatibility of tools, replacement parts, and equipment. Incompatibility between U.S. and British screws prevented the interchanging of the two allies' tank parts in North Africa during World War II, thereby immobilizing significant numbers of vehicles at critical times. This particular problem was rectified in 1948 with the adoption of an international screw thread standard. However, these types of occurrences during the War highlighted the need for standards aimed at reducing inventories and increasing compatibility. Since World War II, the importance of standards has continued to escalate rapidly at both the national and international levels.

TYPES OF STANDARDS

There are two different types of standards -- physical measurement standards and documentary standards. The National Institute of Standards and Technology (NIST) is responsible for developing, maintaining and disseminating national physical measurement standards for basic measurement quantities (such as mass, time and frequency), which are traceable to the International System of Units (SI). While this document does not address physical measurement standards, internationally

---

National Bureau of Standards, U.S. Department of Commerce, Washington, D.C., pp. 82-86, 1974.
[9] American Standards Association, "Through History with Standards" in Rowen Glie (ed.), <u>Speaking of Standards,</u> Cahner Books, Boston, MA, 1972, p. 60.
[10] American Standards Association, p. 60.
[11] Wikepedia, http://www.answers.com/topic/standard-gauge
[12] American National Standards, Ibid, p. 50.

traceable physical measurement standards combined with highly accurate measurements underpin most documentary standards. Together they also promote order, efficiency, and fairness in the marketplace, facilitate technological progress, and enhance U.S. competitiveness.

For simplicity, this report focuses on documentary standards,[13] which are written agreements containing technical specifications or other precise criteria that may contain rules, guidelines, or definitions of characteristics. Standards ensure that materials, products, personnel qualifications, processes, and services are: adequate for their purpose, compatible and/or interchangeable, if necessary; ensure public health and safety; protect the environment; and/or improve economic performance.

Standards can specify product characteristics, establish accepted test methods and procedures, characterize materials, define processes and systems, or specify knowledge, training and competencies for specific tasks. There are numerous ways to classify documentary standards, some of which are described here.

ISO/IEC Guide 2: 2004[14] differentiates eight common types of standards based on purpose. A *basic standard* has a wide-ranging coverage or contains general provisions for one particular field, such as a standard for metal that can affect a wide range of products from cars to fasteners.

Terminology standards are concerned with terms, usually accompanied by their definitions. The standards define words that permit industries or parties entering into a transaction to use a common, clearly understood language.

Testing standards are concerned with test methods, sometimes supplemented with other provisions related to testing, such as sampling, use of statistical methods, or the sequence of tests. They are generally used to assess the performance or other characteristics of a product.

Product standards specify requirements to be fulfilled by a product (or a group of products) to establish its fitness for purpose. Such standards can also address other issues, including packaging and labeling or processing requirements.

Process standards specify requirements to be fulfilled by a process to establish its fitness for purpose. For example, a process standard could cover requirements for the effective functioning of an assembly line operation.

Service standards, such as for servicing or repairing a car, establish requirements to be fulfilled by a service to establish its fitness for purpose.

---

[13] Hereafter, the term "standard" in this report means a documentary standard unless otherwise specified.
[14] ISO/IEC Guide 2: 2004, Standardization and related activities -- General vocabulary, provides general terms and definitions concerning standardization and related activities.

Interface standards, such as requirements for the point of connection between a telephone and a computer terminal, specify requirements concerned with the compatibility of products or systems at their points of interconnection.

Standards on data to be provided contain a list of characteristics for which values or other data are to be stated for specifying the product, process or service. This type of standard generally provides a list of data requirements for a product or service for which values need to be obtained.

It should be noted that these categories are not mutually exclusive. For instance, a product standard may contain testing requirements and therefore also be a testing standard. It may also contain a list of standardized terms and be a terminology standard as well.

Another important type of standard is called a *harmonized standard*. Harmonized standards result from attempts by a nation or a standards developing organization to make its standards compatible with international, regional or other types of standards. A harmonized standard can also result when two or more nations agree on the content and application of a standard. This latter type of harmonized standard tends to be mandatory. It should be noted that in the European standards system, the term "harmonized standard" has a distinctly different meaning and tends to refer to creation of standards that are integrated into the regulatory system within the European Union.[15]

Standards may also be classified by the intended user group or by the standard's developer. There are *company standards*, developed for use by a company or organization for its own products or for the products it purchases. There are also *international standards*, most of which are developed and promulgated by international governmental and nongovernmental organizations, such as the International Telecommunication Union (ITU) (governmental) and ISO (nongovernmental). However, there are other types of international standards. Some company standards gain such widespread marketplace acceptance that they can and do become de facto *international standards,* such as the architecture for the personal computer established by IBM and widely used in the personal computer industry. There are also standards developed by many standards developing organizations that are considered to be international standards because of their global acceptance and usage. Such standards include the American Society of Mechanical Engineers' (ASME) Boiler and Pressure Vessel Code, which is used in at more than 60 countries.[16]

There are also foreign national standards developed by organizations in other countries and regional standards that are developed by a particular group of countries in a geographical region. There are still other classifications such as *industry standards*, developed and promulgated by an industry for materials, products, processes, and services related to that industry. *Government regulatory standards* are those designed to be used by federal regulatory agencies in rulemaking and related activities. These should not be confused with *Federal* and *Military Specifications*, which are official

---

[15] For further information, see NIST SP 951, *A Guide to EU Standards and Conformity Assessment*, by Helen Delaney and Rene van de Zande, which is available at: http://ts.nist.gov/Standards/Global/sp951.cfm.

[16] Domenic Canonico, "A Look at ASME's Boiler and Pressure Vessel Code (BPVC)," from the ASME publication, "ASME Codes and Standards: Examples for Use for Mechanical Engineering Students," ASME, New York, NY, Page 4.

documents used by agencies and by the Department of Defense respectively, to support government procurement. Specifications are a set of conditions and requirements that provide a detailed description of a procedure, process, material, product, or service for use primarily in procurement and manufacturing.[17]

Yet another distinction among standards is the manner in which they specify requirements. Those standards that describe how a product is supposed to function are called *performance standards*. In contrast, *design standards* define characteristics or how the product is to be built. For example, a performance standard for water pipe might set requirements for the pressure per square centimeter that a pipe must withstand, along with a test method to determine if a specimen meets the requirement. On the other hand, the requirement that a pipe must be made of a given gauge of copper would be a design standard.

However, the distinction between these two types of standards is not always clear cut. It is possible to include two different requirements within the same standard -- one of which is stated in terms of performance and the other in terms of design. For example, in a standard for copper pipe, requirements for the pipe can be specified in terms of its performance (being able to withstand a given amount of pressure), but the same standard may require that the pipe's flanges or couplings meet specific design requirements. Few standards are purely design or performance in nature. Most are a mix of requirements of both types. In addition, even if requirements in a standard are mostly written in terms of performance, the test method for verifying or determining conformance is likely to be written in design terms. In fact, design requirements are frequently more appropriate for test methods where the need for accuracy and reproducibility usually outweigh other considerations.

It should also be noted that the determination of conformance to performance standards may be more difficult than for design standards. For example, it is usually more difficult to determine that a pipe can perform in the specified manner than it is to determine that a pipe is made of a given gauge of copper and has a given diameter. Performance standards are also more difficult to write. Therefore, the use of performance standards, while desirable, may not be practical in all situations. In some cases, the disadvantages associated with the use of performance standards may outweigh other considerations.

However, in general, when products can be defined in terms of required performance characteristics, the resulting performance standards tend to be less restrictive than design standards. Performance standards are also more likely to allow the inclusion of technological innovations in the product and to prevent unnecessary barriers to trade.

Yet another classification scheme distinguishes between *voluntary standards*, which by themselves impose no obligations regarding use, and *mandatory standards*. Mandatory standards are set by government regulatory agencies at all levels -- state, local, and federal. They are usually included within the regulations of the government agency with applicable jurisdiction. Such regulations or

---

[17] W. E. Andrus, Jr., Draft NBS Glossary of Terms for Product Standardization, Product Certification and Laboratory Accreditation, U.S. National Bureau of Standards, Dept. of Commerce, 1974.

mandatory "standards,"[18] generally establish requirements for public health and safety, consumer protection, environmental protection, national security, or other similar criteria.

Voluntary standards are generally produced by private sector organizations engaged in the development of standards. Participation in their development is optional and the resulting standards are generally intended for voluntary use.

However, the distinction between these two categories may be lost when voluntary consensus standards are referenced in government regulations, effectively making them "mandatory" standards. Voluntary consensus standards may also become "*quasi-mandatory*" due to conditions in the marketplace. For example, the health care industry is very sensitive to the need to use the safest products available to ensure patient safety and to protect manufacturers, vendors and health care providers against lawsuits. Informed buyers of health care products will frequently insist that products meet all appropriate voluntary consensus standards. If they wish to compete effectively, manufacturers of such products are obliged to conform to such standards in addition to applicable regulations.

It is clear that standards cover a broad range of types and serve a wide variety of purposes. In the United States alone, there are approximately 50,000 private sector voluntary standards[19] developed by more than 600[20] organizations. This number does not include the more than 44,000 distinct statutes, technical regulations or purchasing specifications developed and used by federal regulatory and procurement authorities.[21] It also does not include other codes, rules and regulations containing standards, which have been developed and adopted by state and local government authorities.

Standards are therefore vital tools of industry and commerce because they promote understanding between buyers and sellers and make possible mutually beneficial commercial transactions.

STANDARDS DEVELOPMENT PRINCIPLES

Requirements for the development of standards vary among domestic, regional, international and foreign organizations. However, certain principles for standards development are widely accepted. To maximize their utility and to prevent the creation of unnecessary barriers to foreign and domestic commerce, the standards making process should be conducted in accordance with the internationally

---

[18] This type of regulation is referred to in the World Trade Organization Agreement on Technical Barriers to Trade as a "technical regulation."

[19] American National Standards Institute, "Overview of the U.S. Standardization System," available on ANSI's StandardsPortal.org website at:
http://www.standardsportal.org/usa_en/standards_system.aspx.

[20] Standards & Competitiveness: Coordinating for Results, U.S. Department of Commerce, May 2004, Page 5.

[21] American National Standards Institute, "Overview of the U.S. Standardization System," available on ANSI's StandardsPortal.org website at:
http://www.standardsportal.org/usa_en/standards_system.aspx.

accepted principles of Consensus, Transparency, Balance, Due Process, and Openness. These principles are particularly important for standards likely to be used in technical regulations.

These five principles are explained below:

Consensus. Consensus means that all views are heard and the resultant standard is generally agreed to by those involved. Consensus is characterized by the absence of sustained opposition to substantive issues. However, it does not necessarily imply unanimity.

Transparency. Transparency means: (a) providing advance public notice of a proposed standards development activity; (b) identifying the scope of work to be undertaken; (c) providing information on conditions for participation; (d) and providing an opportunity for all interested parties to comment prior to final approval and adoption.

Balance. Balance means that no one interest, including the government, should dominate. It should be noted that balance can be affected not only by the number of participants in particular categories but also by the funding source. The provider of the funding in standards development work can sometimes end up dominating the process. This is particularly true if the funding is from a government entity. If funding is to be provided by a government agency or other entity, care must be taken to avoid undue influence on the outcome of the process by the funding provider.

Due Process. Due process means that any person (organization, company, government agency, individual, etc.) with a direct and material interest has a right to: (a) express a position and the basis for it; (b) have that position considered; and (c) appeal if adversely affected. Due process ensures equity and fair play in the standards development process.

Openness. The standards development process should be to open to participation by all materially affected interests.

Most of the nation's SDOs operate according to these principles; and the result is an open, competitive system that has produced standards that are widely recognized for the high quality of their technical content. The World Trade Organization's (WTO) Committee of Technical Barriers to Trade (TBT) has recognized the principles listed above, and has recommended the following additional principles to clarify and strengthen the concept of international standards development:[22]

Impartiality. All countries should be provided with meaningful opportunities to contribute to the development of international standards so that the standards development process will not favor the interests of a particular supplier(s), country(ies) or region(s).

Effectiveness and relevance. International standards need to be relevant and effectively respond to regulatory and market needs, as well as scientific and technological developments in various

---

[22] World Trade Organization (WTO) Committee on Technical Barriers to Trade "Decision of the committee on principles for the development of international standards, guides and recommendations with relation to articles 2, 5 and annex 3 of the agreement," G/TBT/1/Rev.8, 23 May 2002.

countries. They should not distort the global market, have adverse effects on fair competition, or stifle innovation and technological development. In addition, they should not give preference to the characteristics or requirements of specific countries or regions when different needs or interests exist in other countries or regions. Whenever possible, international standards should be based on performance rather than on design or descriptive characteristics.

Coherence. To avoid the development of conflicting international standards, it is important that international standards developing organizations avoid duplication of, or overlap with, the work of other international standards developing organizations. Cooperation and coordination with other relevant international organizations are essential.

Development dimension. The ability of developing countries to effectively participate in standards development should be considered in the international standards development process. Tangible ways of facilitating developing countries' participation should be sought.

Compliance with these principles helps standardization activities in the United States and around the world by protecting the rights and interests of participants, while reducing the probability that any resultant standards will become barriers to domestic and international commerce or inhibit the introduction of new technologies.

THE U.S. STANDARDS SYSTEM

As noted in the 1995 National Research Council report on standards and conformity assessment, "The U.S. standards development system serves the national interest well. In most cases, it supports efficient and timely development of product and process standards that meet economic and public interests."[23] Although the exact number is difficult to quantify, it has been estimated that the United States has approximately 50,000 current voluntary standards that have been developed by more than 600 organizations. These do not include an estimated 44,000 distinct statutes, technical regulations or purchasing specifications, developed and used by federal agencies. or the regulations and purchasing specifications containing standards developed and adopted at state and local government levels. There are private and government standards for virtually all industries and product sectors.

Compared to other developed nations, the U.S. standardization structure is highly decentralized. While more than 600 nongovernmental organizations comprise the U.S. standardization system, the situation is not quite as complex as it would appear at first glance. Approximately 19 of these standards developing organizations (SDOs) generate the vast majority of standards in the United States. These SDOs include: ASTM International; Association of American Railroads (AAR); American Association of Cereal Chemists (AACC); American Association of State Highway and Transportation Officials (AASHTO); American Conference of Government Industrial Hygienists (ACGIH); American Oil Chemists Society (AOCS); American Petroleum Institute (API); American Railway Engineers Association (AREA); American Society of Mechanical Engineers (ASME), U.S. Pharmacopia; the Association of Official Analytical Chemists (AOAC); the Cosmetic,

---

[23] National Research Council, Standards, Conformity Assessment, and Trade: Into the 21st Century, National Academies Press, 1995, p. 3.

Toiletry and Fragrance Association (CTFA), now known as the Personal Care Products Council; the Society of Automotive Engineers (SAE); the Aerospace Industries Association (AIA); the Electronic Industries Association (EIA); the Institute of Electrical and Electronics Engineers (IEEE); Technical Association of the Pulp and Paper Industry (TAPPI), Underwriters Laboratories (UL); and the National Fire Protection Association (NFPA). These 19 leading private sector standards developers produce standards that encompass a spectrum of industry sectors, including: aerospace; electronics; automotive and mechanical engineering; petroleum products; chemicals; pulp and paper; and cosmetics. This group also includes developers of safety-related standards, such as those for fire protection, industrial hygiene, consumer product safety, and industrial product safety and protection.

Many of these organizations produce standards that are used globally and encourage participation by foreign technical experts and other interested parties. All foreign participants are considered to be subject matter experts and not national representatives of their government or their national standards body. Comments from both domestic and foreign participants are evaluated and accepted or rejected based on the comments' technical validity.

In some ways, the United States is very different from other countries of the world, where usually one organization is designated as the major standards developer and that organization is closely tied to, if not a part of, the government. The U.S. standards system is primarily voluntary, private sector, and marketplace driven with multiple standards developers taking an active role. Unlike other nations where governments play a more active role and the process is more centralized, the U.S. federal government participates only as one of many stakeholders in the standards development process and not as the driver of the process.

The U.S. system is also tremendously diverse and the result is a system that is largely sectoral in its focus. This is a logical approach because each industrial sector, such as the information technology, telecommunications, automotive, medical devices, and building technology sectors, is most likely to understand that sector's needs and to know what standards best meet those needs. Compared with umbrella-type standards organizations that operate in other nations or at the international level, the more specialized U.S. SDOs also tend be quicker to generate standards needed by industry.

U.S. PRIVATE SECTOR STANDARDS DEVELOPING ORGANIZATIONS (SDOs)

There are many types of organizations that develop standards in the United States, most of which were established in response to a specific marketplace need. The need for safe and economical structures, such as roads and bridges, led to the founding of the International Association for Testing and Materials in 1896. Its mission was to develop standardized test methods. Two years later, the American Section of this organization was formed and became the forerunner of the American Society for Testing and Materials, now known as ASTM International. Since becoming an independent organization in 1902, ASTM has continued to grow and now produces the largest number of nongovernmental, voluntary standards in the United States -- more than 12,000 standards, covering metals, petroleum, construction, the environment, and more.

Another of the major private standards organizations, the American Society of Mechanical Engineering (ASME), was founded in 1880 and first issued the ASME Boiler Code in 1914. The ASME Boiler and Pressure Vessel Code has currently been adopted in part or in its entirety by 49 states and numerous municipalities and territories of the United States and all the provinces of Canada.[24] The Code is also recognized in approximately 60 countries throughout the world.[25] The ASME Boiler Code may be the most widely used voluntary standard in the world.

The founding of the Society of Automotive Engineers (SAE) in 1910 led to the pioneering efforts of the U.S. automotive industry to achieve substantial inter-company technical standardization. Most drivers now take these efforts for granted when choosing motor oils by SAE designations (such as 10W-40) without being aware of the full significance and background of the detailed standards development process.

Most consumers also take for granted the familiar UL mark on a range of products from electrical appliances to fire extinguishers. Founded in 1894, Underwriters Laboratories (UL) is not only a major standards writer, but also operates non-profit testing laboratories and certification programs whose mission is to investigate products and materials with respect to hazards that might affect life or property and to list those items which appear to pose no significant hazards.

The work of other major standards organizations, although equally vital, tends to be less well known outside the standards community. For example, the Institute of Electrical and Electronics Engineers (IEEE), which traces its origin back to 1884, maintains more than 500 standards with 800 more under development. The National Fire Protection Association (NFPA) has for more than three quarters of a century produced the National Electrical Code that is used in building construction. NFPA has also produced many other standards affecting our safety from fires and other hazards. We accept without thought the safety of aircraft -- unaware of the standards produced by the Aerospace Industries Association of America (AIA) for guidance and control systems and other aerospace-related equipment and materials. The Association of American Railroads' (AAR) standards similarly affect the safety and performance of our railroads. Even the quality and size of paper is standardized through the work of the Technical Association of the Pulp and Paper Industry (TAPPI).

There are five main types of U.S., private sector standards developing organizations. The first includes technical and professional societies, such as IEEE and the NSF International (formerly the National Sanitation Foundation), that engage in technical standards development and whose membership is generally composed of individuals who practice a particular profession or discipline. Second are industry associations, such as the National Electrical Manufacturers Association (NEMA), whose membership consists of companies that operate in a specific industry sector. The third group is composed of standards-developing membership organizations, such as ASTM International. The primary focus of these organizations is standards development and standards-related activities, unlike trade associations and technical and professional societies for whom standards development is just one of many activities. The fourth group is composed of building code organizations, such as the International Code Council (ICC). These organizations are composed of building, construction, zoning, and inspection officials. They have developed model building codes

---

[24] From the ASME website at: http://www.asme.org/Codes/International_Boiler_Pressure.cfm.

[25] Domenic Canonico, "A Look at ASME's Boiler and Pressure Vessel Code (BPVC)," from the ASME publication, "ASME Codes and Standards: Examples for Use for Mechanical Engineering Students," ASME, New York, NY, Page 4.

that have been adopted by thousands of State and local governments within the United States.

The last group is composed of non-traditional standards developing organizations, known as consortia.[26] Consortia are groups of like-minded companies and other interested parties who gather together to produce specific standards of interest to the membership. Over the past decade, while there have been a number of consolidations and mergers, the number of standards consortia has grown to over several hundred.[27] Such organizations, which occur primarily in rapidly developing industrial sectors, were established to serve as a more rapid forum for standards development than the more formal processes of traditional standards developing organizations. Time pressures on standards development have increased in many sectors because new products, such as those in the information and communication technologies (ICT) sector, have short shelf lives and must be brought to market quickly. Traditional SDO processes are often slower to meet industry needs in such areas.

Consortia activities generally operate on a pay-to-play membership basis, and their standards are often free. The degree to which consortia meet ANSI's criteria for consensus, openness, balance, due process, transparency, varies among individual consortia. At one extreme are organizations, such as the Internet Engineering Task Force (IETF), which is operated in a very open and transparent fashion with membership that is open to all interested parties. At the other extreme are organizations whose meetings are generally closed and whose membership is limited to companies in a specific industry.

In addition, some consortia have an affiliation with more traditional standards developers, such as the IEEE's Industry Standards and Technology Organization (IEEE-ISTO). Others operate totally independently. The broad range of organizations participating in standards development reflects the impact standards have on a vast spectrum of interests and disciplines.

ROLE OF THE NATIONAL INSTITUTE OF STANDARDS AND TECHNOLOGY (NIST)

NIST is a federal agency within the U.S. Department of Commerce whose mission is to promote U.S. innovation and industrial competitiveness by advancing measurement science, standards, and technology in ways that enhance economic security and improve the quality of life. It was established in 1901 by an act of Congress as the U.S. measurement institute. NIST has approximately 2,900 employees, 2,600 associates and facility users, and 1,600 affiliated field agents located in Gaithersburg, Maryland and Boulder, Colorado. The measurements, standards, and technologies that are the essence of the work done by NIST's laboratories help U.S. industry and researchers to invent and manufacture superior products and to provide services reliably. In addition, NIST manages some of the world's most specialized measurement facilities in the country.[28]

---

[26] For those interested in the subject of standards consortia, a wealth of information is available from Consortium Info.org on its website at: http://consortiuminfo.org.

[27] Carl F. Cargill, "Consortia Standards: Towards a Re-Definition of a Voluntary Consensus Standards Development Organization," Testimony before the Subcommittee on Environment, Technology, and Standards; Committee on Science, U.S. House of Representatives, June 28, 2001, p.2.

[28] For additional information, see:
http://www.nist.gov/public_affairs/factsheet/strengthen_economy_safety.htm

In addition to its other responsibilities, NIST has a variety of roles in the private sector-led U.S. voluntary standards system. As the national measurement institute, NIST is frequently looked to for research and measurements that provide the technical underpinning for standards, ranging from materials test methods to standards for building performance, and for a range of technologies, from information and communications technologies to nano- and bio-technologies. NIST staff frequently participate in the preparation of the standards documents themselves, typically through their work on private sector-led standards committees. NIST staff also participate in workshops, seminars, and conferences supporting these standards activities.

Under the National Technology Transfer and Advancement Act (NTTAA), NIST was also given responsibility for coordinating federal, state and local activities in voluntary standards and working with industry and government to develop and apply technology, measurements and standards. In addition, NIST is responsible for chairing the Interagency Committee on Standards Policy (ICSP), which helps to ensure effective participation by the federal government in domestic and international standards and conformity assessment activities and promote the adherence to uniform policies by federal agencies in the development and use of standards and in conformity assessment activities.

In addition, NIST is responsible for:
- Operating the U.S. National Inquiry Point on Technical Barriers to Trade, which provides research services on standards, technical regulations, and conformity assessment procedures for non-agricultural products to assist in carrying out the U.S. government's responsibilities under the World Trade Organization (WTO) Agreement on Technical Barriers to Trade (TBT);
- Operating the National Voluntary Laboratory Accreditation Program (NVLAP), which provides third-party accreditation to testing and calibration laboratories in response to Congressional mandates or administrative actions by the Federal Government or from requests by private-sector organizations and operates in conformance with International Organization for Standardization (ISO) and the International Electrotechnical Commission (IEC) standards, including ISO/IEC 17025 and ISO/IEC 17011.
- Maintaining the fundamental physical standards, such as length, time and frequency and units of mass, which underlie measurements contained in standards.

To more effectively coordinate its standards role with that of the private sector, NIST has also entered into a Memorandum of Understanding (MOU) with the American National Standards Institute (ANSI).[29] The MoU is intended to improve domestic communication and coordination among both private and public sector parties in the United States on voluntary standards issues and increase the effectiveness of U.S. government agency participation in the national and international voluntary standards-setting process.

ROLE OF THE AMERICAN NATIONAL STANDARDS INSTITUTE (ANSI)

---

[29] For a copy of the MoU, see http://ts.nist.gov/Standards/Conformity/ansimou.cfm

ANSI has served as administrator and coordinator of the United States private sector, voluntary standardization system for almost 90 years. Founded in 1918 by five engineering societies and three government agencies, the ANSI remains a private, not-for-profit membership organization supported by a diverse constituency of private and public sector organizations. The Institute is comprised of government agencies, organizations, companies, academic and international bodies, and individuals. ANSI represents the interests of nearly 125,000 companies and 3.5 million professionals through its office in New York City and its headquarters in Washington, DC.

Among its standards-related activities, ANSI accredits U.S. standards developers using criteria based on international requirements.[30] ANSI has accredited over 200 standards developers in the private and public sectors. These accredited SDOs develop standards based on consensus and other principles, and can choose to publish such standards as American National Standards (ANS). At the end of 2003, there were more than 10,000 such documents.[31]

Due process is the key to ensuring that ANSs are developed in an environment that is equitable, accessible and responsive to the requirements of various stakeholders. Furthermore, ANSI accreditation assures the accredited SDOs follow an open and fair process where all interested and affected parties have an opportunity to participate in a standard's development and to have their views considered.

ANSI is the sole U.S. representative and dues-paying member of the two major non-treaty international standards developing organizations, the ISO; and, via the U.S. National Committee (USNC), the International Electrotechnical Commission (IEC). Through ANSI, the U.S. has immediate access to the ISO and IEC standards development processes. ANSI participates in almost the entire technical program of both the ISO and the IEC and administers many key committees and subcommittees. Part of its responsibilities as the U.S. member body to the ISO includes accrediting U.S. Technical Advisory Groups (U.S. TAGs). The primary purpose of U.S. TAGs is to develop and transmit U.S. positions on ISO and IEC activities and ballots via ANSI or the USNC Technical Management Committee (TMC).

In many instances, U.S. standards are taken forward to ISO and IEC, through ANSI or the USNC, where they are considered and often adopted in whole or in part as international standards. Through this mechanism, ANSI plays an important role in creating international standards that support global commerce and which can prevent or discourage countries from developing and/or adopting local standards that favor their domestic industries and create barriers to international trade. Since volunteers from industry and government, not ANSI staff, carry out the work of the international technical committees, the success of these efforts is often dependent on the willingness of U.S. industry and government to commit the resources required to ensure strong U.S. technical participation in the international standards process.

---

[30] Effective 2003-2004, the *ANSI Essential Requirements* replaced the *ANSI Procedures for the Development and Coordination of Standards* as ANSI accreditation criteria for standards developers.

[31] American National Standards Institute, "Domestic Programs (American National Standards) Overview," http://www.ansi.org/standards_activities/domestic_programs/overview.aspx?menuid=3

In December 2000, NIST and ANSI renewed their Memorandum of Understanding (MOU), which outlines the role of each organization and provides the basis for ongoing, cooperative efforts to enhance and strengthen the U.S. voluntary, consensus standards system and to support continued U.S. competitiveness, economic growth, health, safety, and protection of the environment through strong public-private sector partnership. The ANSI-NIST MOU outlines the roles of each organization and provides the basis for positive ongoing cooperative efforts. The MOU has been particularly useful in coordinating the activities of federal agencies in their transition to greater use of voluntary consensus standards as the result of the passage of the National Technology Transfer and Advancement Act (NTTAA).

THE INTERNATIONAL STANDARDS SYSTEM

There are also numerous international organizations that produce standards. Some are operated within the private sector, while others are governmental organizations established by treaty. There are also standards developing organizations, some of whose standards are considered to be international because of their global usage. In addition, there are a growing number of consortia that operate globally and outside of the more traditional standards system.

Among the private sector bodies, the International Organization for Standardization (ISO) is probably the largest producer of International Standards, having issued over 16,000 standards. ISO's work is carried out through some 3,000 technical groups in which experts from roughly 157 countries participate annually. [32]

There are also other international, private sector standards developing organizations, such as the International Electrotechnical Commission (IEC) that develops standards in the electrical and electronic area and operates in a manner similar to its sister organization, ISO.

In addition there are treaty (governmental) organizations in which the official representative is a government entity. Many of these organizations allow, and even encourage, participation by relevant private sector entities in their discussions. A few of these treaty organizations are discussed below.

The Codex Alimentarius Commission (Codex or CAC) develops food safety standards; the World Health Organization (WHO) develops health-related standards; and the International Telecommunications Union (ITU) develops standards in the radio and telecommunications area.

The World Meteorological Organization (WMO)[33] is an intergovernmental organization established in 1950. WMO is the specialized agency of the United Nations for meteorology (weather and climate), operational hydrology and related geophysical sciences, and is responsible for (among other goals) promoting standardization of meteorological and related observations and ensuring the uniform publication of observations and statistics.

---

[32] A description of the ISO standards development process is available at: http://www.iso.org/iso/standards_development/processes_and_procedures.htm

[33] For additional information on WMO, see: http://www.wmo.ch/pages/index_en.html.

Established in 1874, the Universal Postal Union (UPU)[34] is the primary forum for cooperation between postal-sector players. With 192 member countries, this specialized agency of the United Nations sets the standards and rules for international mail exchanges and makes recommendations to stimulate growth in mail volumes and to improve the quality of service for customers. Standards are important prerequisites for effective postal operations and for interconnecting the global postal network. This information is published and available to the general public. To date, over 100 technical standards have been developed by the UPU. The UPU's open approach to the development of postal standards allows all stakeholders in the postal industry to participate actively in the standards development process.

The World Customs Organization (WCO)[35] is responsible for the development of the Harmonized Commodity Description and Coding System, generally referred to as "Harmonized System" or simply "HS." The system is used by more than 190 countries and economies as a basis for their Customs tariffs and for the collection of international trade statistics. Over 98 % of the merchandise in international trade is classified in terms of the HS.

There are also organizations such as the North Atlantic Treaty Organization (NATO),[36] which develops Standards Agreements (STANAGs) and related publications for use by NATO member countries.

As previously mentioned, there are also a number of U.S.-based organizations that produce standards that are adopted and used internationally, such as ASTM International and ASME. These organizations are generally open to participation by any interested national or foreign person or organization, and the standards they produce are used globally.

There are also many consortia[37] that operate in the global arena and are open to membership from all countries. Two such organizations, the Internet Engineering Task Force (IETF) and the World Wide Web Consortium (W3C),[38] are considered to be primarily responsible for the standards required for the development of the Internet.

The Third Generation Partnership Project 2 (3GPP2)[39] is a collaborative third generation (3G) telecommunications specifications-setting project comprised of North American and Asian interests. It is responsible for developing global specifications for ANSI/Telecommunications Industry Association (TIA)/Electronic Industries Association (EIA)-41 Cellular Radiotelecommunication Intersystem Operations network evolution to 3G and global specifications for the radio transmission technologies (RTTs) supported by ANSI/TIA/EIA-41. 3GPP2 is a collaborative effort between five officially recognized SDOs, including the Telecommunications Industry Association (TIA).

---

[34] For additional information on UPU, see http://www.upu.int/about_us/en/upu_at_a_glance.html.
[35] For additional information on WCO, see http://www.wcoomd.org.
[36] See: http://www.nato.int/cps/en/natolive/stanag.htm for a list of such agreements and publications.
[37] For those interested in the subject of standards consortia, a wealth of information is available from Consortium Info.org on its website at: http://consortiuminfo.org.
[38] For additional information on W3C, see: http://www.w3.org.
[39] For additional information on 3gpp2, see: http://www.3gpp2.org.

In addition, the Automotive Open System Architecture (AUTOSAR)[40] is responsible for an open, standardized automotive software architecture, jointly developed by automobile manufacturers, suppliers and tool developers. Accellera's[41] mission is to encourage the worldwide development and use of standards required by systems, semiconductor and design tools companies, which enhance a language-based design automation process. These are just a few of the hundreds of such consortia that operate globally.

Like the U.S. standards system, the international standards system reflects a diversity of interests, organizational types, and scopes.

REGIONAL STANDARDS SYSTEMS

In addition to international organizations, there are also many regional standards organizations. The standards of regional organizations are generally developed by and relate to only one region of the world, though such bodies can and often do promote the use of their standards in other regions. Some regional organizations are governmental and treaty based, while others are operated by the private sector. A few of the major regional standards organizations are described below.

The African Regional Organization for Standardization (ARSO)[42] serves the interests of the national standards bodies of: Burkina Faso, Cameroon, Congo Brazzaville, Egypt, Ethiopia, Gabon, Ghana, Kenya, Libyan Arab Jamahiriya, Madagascar, Malawi, Mauritius, Nigeria, Democratic Republic of the Congo, Republic of Guinea, Republic of Senegal, Rwanda, South Africa, Sudan, Tanzania, Tunisia, Uganda, and Zimbabwe. ARSO has developed 733 standards. However, in the future, ARSO will work on the harmonization of national standards and sub-regional standards as African standards and will no longer develop standards.

The Southern African Development Community's (SADC) Cooperation in Standardization (SADCSTAN)[43] is part of SADC and is open to membership from any SADC member body (Angola, Botswana, the Democratic Republic of Congo, Lesotho, Madagascar, Malawi, Mauritius, Mozambique, Namibia, South Africa, Swaziland, United Republic of Tanzania, Zambia and Zimbabwe). The harmonization of standards and technical regulations of member countries is the responsibility of SADCSTAN.

The Arab Industrial Development and Mining Organization[44] (AIDMO), is a specialized governmental organization of the Arab League of States, which aims to establish Arab unified standards and rules of origin of Arab Industrial Commodities for the Arab free trade zone. AIDMO also promotes technical, technological and industrial co-operation among Arab states and with foreign countries. AIDMO members include: Jordan, United Arab Emirates, Bahrain, Tunisia, Algeria, Djibouti, Saudi Arabia, Sudan, Syria, Somalia, Iraq, Oman, Palestine, Qatar, Kuwait,

---

[40] For additional information on AUTOSAR, see: http://www.autosar.org
[41] For additional information on Accellera, see: http://www.accellera.org/home.
[42] For additional information on ARSO, see: http://www.arso-oran.org.
[43] For additional information on SADCSTAN, see: http://www.sadcstan.co.za.
[44] For additional information on AIDMO, see: http://www.aidmo.org.

Lebanon, Libya, Egypt, Morocco, Mauritania, and Yemen.

The Asia Pacific Economic Cooperation (APEC)[45] was established in 1989 to enhance economic growth and prosperity in the Asia-Pacific region. The APEC Sub-Committee on Standards and Conformance (SCSC) works in the field of standards and conformity assessment and encourages greater alignment of the standards of APEC member countries with international standards. The SCSC has also implemented a Mutual Recognition Arrangement (MRA) for conformity assessment of electrical and electronic equipments (EE MRA) as a voluntary scheme to help accelerate a region-wide MRA and is working on other MRAs.

The Interstate Council for Standardization, Metrology and Certification serves the Commonwealth of Independent States (CIS), whose membership includes: Armenia, Azerbaijan, Belarus, Georgia, Kazakhstan, Kyrgyzstan, Moldova, Russia, Tajikistan, Turkmenistan, Ukraine, and Uzbekistan. The Interstate Council is the CIS intergovernmental body for the formulation and implementation of coordinated policy in the field of standardization, metrology and certification. The Interstate Council's working body is the Bureau for Standards that comprised of groups of experts and the Regional Information Center. More than 230 interstate technical committees for standardization exist under the Council. The Interstate Council is recognized by ISO under the name, the EuroAsian Council for Standardization, Metrology and Certification (EASC). [46]

The Organization for Economic Cooperation and Development (OECD)[47] is a group of 30 member countries, including the United States, that share a commitment to democratic government and the market economy. Discussions at the OECD committee level sometimes evolve into negotiations where OECD countries agree on rules for international cooperation. They can culminate in formal agreements by countries (e.g., combating bribery) on arrangements for export credits, or on the treatment of capital movements. They may produce standards and models (e.g., the application of bilateral treaties on taxation) or recommendations (e.g., cross-border co-operation in enforcing laws against spam). They may also result in guidelines (e.g., corporate governance or environmental practices). In addition, OECD has developed schemes for the application of international standards for fruit and vegetables, the official testing of agricultural and forestry tractors, the varietal certification or control of seed moving in international trade, the mutual acceptance of data in the assessment of chemicals, and the control of forest reproductive material moving in international trade

The Andean Community (CAN), a governmental common market, is composed of Colombia, Ecuador, Bolivia, and Peru,[48] has established the Andean Standardization, Accreditation, Testing, Certification, Technical Regulations and Metrology System, approved by Decisions 376 and 419 of the Andean Community. These decisions prescribe the obligation of Member Countries to notify any proposals involving new Technical Regulations, Mandatory Technical Standards, compliance evaluation procedures, mandatory certifications and any other equivalent mandatory measures. The Community is also working towards the harmonization of technical regulations or mandatory

---

[45] For additional information on APEC, see: http://www.apecsec.org.sg/apec.html.
[46] For further information on EASC, see: http://www.easc.org.by/english/mgs_org_en.php.
[47] For additional information on OECD, see: http://www.oecd.org.
[48] For additional information on the Andean Community, see: http://www.comunidadandina.org/endex.htm.

standards.

The Caribbean Common Market's (CARICOM) Regional Organization for Standards and Quality (CROSQ)[49] was established in 2003 by a CARICOM Community treaty as an intergovernmental organization for promoting efficiency and competitive production in trade and services through the process of standardization and the verification of quality. CROSQ, the successor to the Caribbean Common Market Standards Council (CCMSC), is mandated to represent the interests of the region in international and hemispheric standards work, to promote the harmonization of metrology systems and standards, and to increase the pace of standards development in the region. CROSQ also serves as the Regional Accreditation Body and as an inquiry point for the World Trade Organization (WTO).

Europe is served by one governmental body, the UN Economic Commission for Europe (UNECE), which also includes countries outside of Europe. In addition, Europe has three private sector regional bodies:[50] the European Committee for Standardization (CEN), the European Committee for Electrotechnical Standardization (CENELEC) and the European Telecommunications Standards Institute (ETSI). These 3 bodies are collectively called European Standards Organizations (ESOs).

Founded in 1961, CEN is a private, not-for-profit organization, composed of the national standards bodies in the European Union (EU)[51] and the European Free Trade Association (EFTA).[52] CEN develops standards in all fields except the electrotechnical area. In addition to its market driven standards efforts, CEN also develops standards in support of EU and EFTA governmental policies and regulations (directives) in response to formal governmental "mandates." Under the so-called New Approach, EU regulations (directives) are limited to establishing "essential requirements." These essential requirements are obligatory and are formulated in general terms. The detailed technical specifications necessary for the implementation of directives are entrusted to European, voluntary standards organizations like CEN. While the resulting standards are not mandatory, products manufactured according to such "harmonized standards" give a "presumption of conformity" to the essential legal requirements in the directives. Compliance to the directives results in the product's right to bear the CE marking of conformity and to market the product throughout Europe.

CENELEC operates in the electrotechnical area. Created in 1973, CENELEC is a non-profit technical organization composed of the National Electrotechnical Committees of 30 European countries. In addition, eight Affiliate National Committees from neighboring countries participate in CENELEC's work. Like CEN, CENELEC creates both standards requested by the market and

---

[49] For additional information on CROSQ, see: http://www.crosq.org.
[50] For further information on CEN, see: http://www.cen.eu/cenorm/homepage.htm. For further information on CENELEC see: http://www.cenelec.eu/Cenelec/Homepage.htm For further information on ETSI, see: http://www.etsi.org.
[51] The EU is comprised of: Austria, Belgium, Bulgaria, Cyprus, Czech Republic, Denmark, Estonia, Finland, France, Germany, Greece, Hungary, Ireland, Italy, Latvia, Lithuania, Luxembourg, Malta, Netherlands, Poland, Portugal, Romania, Slovakia, Slovenia, Spain, Sweden, and the United Kingdom. Candidates for membership include: Croatia, Macedonia, and Turkey.
[52] EFTA countries include: Iceland, Liechtenstein, Norway and Switzerland.

"harmonized standards" in support of European regulation.

ETSI is an independent, not-for-profit organization that is officially responsible for standardization of information and communication technologies (ICT) within Europe. These technologies include telecommunications, broadcasting and related areas such as intelligent transportation and medical electronics. ETSI includes 655 members from 59 countries inside and outside Europe, including manufacturers, network operators, administrations, service providers, research bodies and users. ETSI is also recognized by the European Commission and the EFTA secretariat.

The Pan American Standards Commission (COPANT)[53] is a private, non-profit association that promotes standardization and related activities for its member bodies in the region of the Americas. COPANT may develop regional standards in limited areas for the specific, regional interests of its members. Where appropriate, COPANT may also adopt international standards. COPANT's active member bodies include: Argentina (IRAM), Barbados (BNSI), Bolivia (IBNORCA), Brazil (ABNT), Canada (SCC), Chile (INN), Colombia (ICONTEC), Costa Rica (INTECO), Cuba (NC), Ecuador (INEN), El Salvador (CONACYT), Grenada (GDBS), Guatemala (COGUANOR), Guyana (GNBS), Honduras (COHCIT), Jamaica (BSJ), Mexico (DGN), Nicaragua (MIFIC), Panama (COPANIT), Paraguay (INTN), Peru (INDECOPI), Dominican Republic (DIGENOR), Saint Lucia (SLBS), Trinidad and Tobago (TTBS), USA (ANSI), Uruguay (UNIT), and Venezuela (FONDONORMA). Adherent members include: the Interamerican Accreditation Cooperation (IAAC), Spain (AENOR), France (AFNOR), Italy (UNI), Portugal (IPQ), and South Africa (SABS).

The Association of Southeast Asian Nations' (ASEAN)[54] Consultative Committee on Standards and Quality (ACCSQ) endeavors to harmonize national standards with international standards and implement mutual recognition arrangements on conformity assessment to achieve its end-goal of "One Standard, One Test, Accepted Everywhere." ASEAN members include: Brunei Darussalam, Cambodia, Indonesia, Laos, Malaysia, Myanmar, the Philippines, Singapore, Thailand and Viet Nam. ACCSQ has conducted considerable work on developing a Mutual Recognition Arrangement (MRAs) for Electrical and Electronic that has been signed by ten member countries with ongoing work toward harmonization of regulatory regimes in the electrical and electronic sector. Other MRA efforts include an agreement in the cosmetics area and one for pharmaceuticals. There are also ASEAN Common Technical Requirements (ATCRs), covering quality, safety and efficacy that are being developed.

There are also a number of other smaller regional organizations, including the Mercado Común del Sur's (MERCOSUR) Standardization Association (AMN),[55] a non-profit, private sector association responsible for the voluntary standardization management within the governmental common market, MERCOSUR. The MERCOSUR Standardization Association is formed by the National Standardization Bodies of the countries that are members: Argentina, Brazil, Paraguay and Uruguay. Bolivia and Chile are associate members.

While these and other regional standards organizations do set standards-related policies and/or

---

[53] For additional information on COPANT, see: http://www.copant.org.
[54] For additional information on ASEAN, see: http://www.aseansec.org.
[55] For additional information on MERCOSUR's AMN, see: http://www.amn.org.br/br.

standards for a specific region that can help to facilitate trade within a defined region, it is difficult to track the work of such organizations because of their sheer number. In addition, if standards are set without regard to standards that have been adopted and used in other parts of the world, barriers to trade can result.

STANDARDS USAGE/CONFORMITY ASSESSMENT[56]

It is important to remember that standards in themselves have little or no significance -- unless and until they are adopted and used. Some standards never receive widespread acceptance and use. Others may have been accepted by industry at one time, but now apply to technologies that have become outdated.

One of the most important uses for standards is within a conformity assessment process. As noted before, buyers cannot always evaluate product specifications or characteristics by inspection or even from prior experience. Information on a product's conformance (or nonconformance) to a particular standard can provide an efficient method of conveying information needed by a buyer on the product's safety and suitability. Standards therefore provide the basis for conformity assessment activities that, in turn, are the basis for many buyer-seller transactions. Hence, standards used in conformity assessment activities can have tremendous impact on companies, nations and the global marketplace.

Standards can cover many aspects of the conformity assessment process. They can describe characteristics of the product for which conformity is sought; the methodology (e.g., test, inspection or other assessment methods) used to assess that conformity; or even the conformity assessment process itself (e.g., how a certification program or conformity assessment body should be operated). Standards used in conformity assessment should be clearly and concisely written, readily understood, precise, technically credible, and contain only unambiguous requirements - the absence or presence of which can be objectively verified. The use of well written standards in a conformity assessment process lends credibility and validity to the process, increasing its usefulness.

In addition, standards used in conformity assessment should not impede innovation. For this reason, performance standards are preferred over design standards. For example, a performance standard for water pipe might set requirements for the pressure per unit area that a pipe must withstand along with a test method to determine if a pipe sample meets the requirement. Manufacturers are free to choose

---

[56] Conformity assessment includes any activity concerned with determining directly or indirectly that requirements for products, services, systems, personnel qualifications and organizations are fulfilled. Conformity assessment includes: sampling and testing; inspection; supplier's declaration of conformity; certification; and management system assessment and registration. It also includes accreditation of laboratories, certifiers, inspection bodies, and management system registrars, and the recognition of the competence of accreditation bodies. Conformity assessment activities may be conducted by the manufacturer/supplier (first party), by the buyer (second party) either directly or by another party acting on the supplier's or buyer's behalf, or by a body not under the control or influence of either the buyer or the seller (third party). It can also be conducted by a government agency acting in a regulatory capacity.

any product design, material, and manufacturing process as long as the pipe can perform in the specified manner. On the other hand, a standard that requires that a pipe be made of a given gauge of copper and have a given diameter is a design standard. Manufacturers trying to comply with such as standard are not free to innovate -- they cannot make the pipe out of stainless steel or some other new material or vary the size of the diameter, even if such changes might improve the pipe's performance. However, it should also be noted that a poorly written standard of *either* type is unlikely to lead to greater technological innovation, increased trade, or to an acceptable outcome when used in a conformity assessment process.

Standards used in conformity assessment should also specify all essential characteristics of a product necessary for achieving the objective of the conformity assessment activity. Knowing what aspects of the product will be evaluated in a conformity assessment process and whether there are other aspects which might impact quality, safety, or performance allows the user of the conformity assessment data to evaluate the data's significance.

In addition, the user must know what standard(s) was used. Given the large number of national, regional, and international standards, it is not surprising that a number of standards are redundant or overlapping. Requirements in two different standards covering the same characteristics may be very different, and different test methods can produce very different results.

Ideally, all standards within a conformity assessment system should be performance based, technically sound and implementable in a cost effective manner.

BENEFITS AND PROBLEMS OF STANDARDIZATION
.
On the whole, the benefits of standardization far outweigh the difficulties and potential for abuse. Standards promote understanding between buyer and seller and make possible mutually beneficial commercial transactions. A product's conformance to accepted standards readily provides an efficient method of conveying complex information on the product's suitability. Architects use standards in a shorthand manner when drafting plans for buildings, and purchasing agents can use standards as an easy way of communicating their needs to potential suppliers. In a host of situations standards are or may be used to replace large quantities of complex information.

Standards underlie mass production methods and processes. They promote more effective and organized social interaction, such as the example of the standardized colors for traffic lights and many other widely accepted conventions. Standards are essential in efforts to improve product safety and to clean up the environment. Standardized and interchangeable parts can reduce inventory requirements and facilitate product repairs. They can also promote fair competition by facilitating the comparison of prices of standardized commodities.

In general, standards permit society to make more effective use of its resources and allow more effective communication among all parties to particular activities, transactions, or processes. Indeed, standards are crucial to every form of scientific and industrial process. Without standards, the quality of life would be significantly reduced.

No system, particularly one as complex and diverse as the U.S. voluntary standards system, is without problems. In a court case of great significance, the United States Supreme Court on May 17, 1982, rendered its decision in favor of Hydrolevel, a manufacturer of low-water fuel cutoff devices, in the case of the American Society of Mechanical Engineers (ASME) v. Hydrolevel. It found ASME liable for conspiring to restrain trade since two subcommittee officers, serving as volunteers but acting in the name of ASME, issued a misinterpretation of a standard and produced an adverse effect on the competitiveness of the plaintiff. Similarly, the Federal Trade Commission held hearings on standards and certification and uncovered "substantiated complaints of individual standards and certification actions that have, in fact, unreasonably restrained trade or deceived or otherwise injured consumers."[57] However, it should also be noted that such cases and events have spurred most U.S. standards developers into enacting polices and procedures and taking aggressive action designed to prevent the recurrence of such problems.

Financial issues associated with development of standards, including both financial support for standards developing organizations and industry sector funding of its participation, are also a concern. The sometimes substantial costs involved in participation in standards development makes it difficult (if not impossible) for small firms and non-industry representatives to be active in the process. In addition, the cost of participation in international standards development can be especially high, and the growth in regional standards development activities further increases demands on limited industry resources. In a number of cases, economic problems have forced even larger companies to cut back on participation. In addition, ensuring adequate consumer representation can be a particular problem. In addition to funding issues, some standards are highly technical in nature. Without sufficient technical expertise, consumers are unlikely to be able to provide meaningful input into the process. Such issues can complicate attempts to achieve balanced representation by all interests concerned.

Other problems can occur when a standard undergoes review and revision. Unless the original technical experts that developed the standard participate in its revision, the reviewers may not be able to fully understand how the document was prepared, what was eliminated from consideration, and the reasons or assumptions underlying decisions and the resultant provisions. Problems can also occur in the application of specific provisions if the intent behind them is unclear. Rationale statements, which sometimes accompany a standard, are specifically designed to define the purpose and scope of the standard, to explain the criteria used in developing its requirements and to provide all other relevant information at the disposal of the developers.[58]

Problems can also occur when standards are not based on sound science. According to the National Foreign Trade Council's May 2003 report, *Looking Behind the Curtain: The Growth of Trade Barriers*, "... when regulations and standards are not based on sound science they serve as de facto trade barriers and have a negative impact on a wide variety of U.S. export sectors, as well as, those of developing countries." The report notes that many standards (or regulations that reference or

---

[57] Bureau of Consumer Protection, Standards and Certification: Final Staff Report - April 1983, Federal Trade Commission, Washington, D.C., April, 1983, p. 2.

[58] David A. Swankin, Rationale Statements for Voluntary Standards—Issues, Techniques, and Consequences, National Bureau of Standards, Dept. of Commerce, Gaithersburg, MD, November, 1981.

incorporate standards) that are not based on sound science and that justify denying market access to imported products on the basis of meeting a national objective (such as the preservation of health and safety, animal welfare and the environment or the protection of consumer choice) may actually be intended to protect ailing or otherwise noncompetitive domestic industries. This study notes that such measures are often based on a "presumption of harm" without any scientific evidence and/or scientifically based risk assessment to support such an assumption. The report notes that such countries invoke "the precautionary principle, a non-scientific touchstone," to justify their enactment of such technical measures and that such an approach is both insular and presumptive of the existence of unacceptable hazard or risk, even in the face of scientific evidence to the contrary. The United States has taken an approach that stresses sound science, risk analysis and transparency. The ability to create a free and open global marketplace depends on the implementation of standards and regulations that are transparent and reference objective principles of sound science.

There can also be problems in s standards development activities when the content of standards ends up being regulatory driven as opposed to marketplace driven. Such standards may not effectively meet the needs of the marketplace, and may also end up as technical barriers to trade.

NATIONAL OBLIGATIONS WITH RESPECT TO STANDARDS

Perhaps the most important piece of U.S. legislation affecting the U.S. standardization system is the National Technology Transfer and Advancement Act (NTTAA),[59] which became law in March 1996. The NTTAA directs U.S. federal agencies on their use of private sector standards and conformity assessment practices. The objective is for U.S. federal agencies to adopt private sector consensus standards wherever possible, in lieu of creating government unique standards. The Act also directs the NIST to bring together U.S. federal agencies, as well as state and local governments, to achieve greater reliance on voluntary standards.

Further guidance on implementing the NTTAA is contained in the Office of Management and Budget's (OMB) Circular A-119, Federal Participation in the Development and Use of Voluntary Consensus Standards and in Conformity Assessment Activities.[60] This Circular directs agencies to use voluntary consensus standards in lieu of government-unique standards except where inconsistent with law or otherwise impractical. It also provides guidance for agencies participating in voluntary consensus standards bodies and describes procedures for satisfying the reporting requirements in the NTTAA. The aim of the Circular is to reduce to a minimum the reliance by agencies on government-unique standards.

There are also other policies and legislation that affect standards adoption and use by specific federal agencies. Such policies and legislation include:[61]

---

[59] The full test of the Act is available at: http://standards.gov/standards_gov/index.cfm?do=documents.NTTAA.

[60] The full text of the Circular is available at: http://standards.gov/standards_gov/index.cfm?do=documents.A119.

[61] American National Standards Institute, "Significant Federal Laws and Policies," http://www.ansi.org/government_affairs/laws_policies/laws.aspx?menuid=6

- The Consumer Product Safety Act, which directs the Consumer Product Safety Commission to rely on voluntary consensus consumer product safety standards rather than promulgate its own standards;
- The Health Insurance Portability and Accountability Act of 1995 that requires the Secretary of Health and Human Services to adopt standards developed by ANSI-accredited standards developers whenever possible;
- The Telecommunications Act of 1996, which contains several provisions that encourage Federal Communications Commission (FCC) reliance on private sector standards;
- The Food and Drug Administration (FDA) Modernization Act of 1997, which contains provisions that allow the FDA in some instances to accept manufacturers' declarations of compliance to certain standards during the evaluation of premarket submissions for electrical medical devices; and
- MILSPEC Reform that has resulted in the Department of Defense's (DoD's) moving away from unique specifications and standards (MILSPECS) and toward reliance on private sector standards.

Such legislation and policies set requirements and goals regarding federal usage of standards.

INTERNATIONAL REQUIREMENTS WITH RESPECT TO STANDARDS

The World Trade Organization (WTO) agreements,[62] to which the United States government is a signatory, contain legal texts that form the foundation and rules for much of the multilateral trading system. One agreement, The Agreement on Technical Barriers to Trade (TBT Agreement), recognizes the important contribution that international standards and conformity assessment systems can make in improving production efficiency and facilitating international trade. This Agreement seeks to ensure that regulations and standards, as well as testing and certification procedures, do not create unnecessary obstacles to trade. The TBT Agreement encourages countries to use international standards where appropriate, but the Agreement does not require countries to change the levels of protection that they consider appropriate. The Agreement also covers processing and production methods related to the characteristics of the product itself.

The TBT Agreement notes that as tariffs have fallen, the number of technical regulations and standards adopted by countries has grown significantly. The Agreement recognizes that a risk exists that technical regulations and standards could be adopted and applied solely to protect domestic industries. To address this problem, the Agreement contains rules for preparation, adoption and application of regulations, standards and conformity assessment procedures. It also encourages the use of performance rather than design standards and regulations where feasible. It notes that unnecessary obstacles to trade can result when a regulation is more restrictive than necessary to achieve a given policy objective, or when it does not fulfill a legitimate objective, such as national security requirements, prevention of deceptive practices, protection of human health or safety,

---

[62] Information on requirements of the WTO Agreement on Technical Barriers to Trade was taken from "Technical Barriers To Trade: Technical Explanation" on the WTO website at: http://www.wto.org/english/tratop_e/tbt_e/tbt_info_e.htm.

protection of animal and plant life or health or the environment.

Some of the provisions of the TBT Agreement related to conformity assessment include:
- avoiding the imposition of stricter or more time-consuming procedures than necessary to assess that a product complies with the domestic laws and regulations;
- limiting information requirements to those necessary to assess compliance;
- siting facilities that carry out conformity assessment and selecting samples in a manner that does not create unnecessary inconvenience;
- treating products imported from the territory of any signatory to the Agreement in no less favorable manner than like products of national origin and like products originating in any other country ("national treatment");
- treating products equally with respect to any fees charged to assess conformity with regulations;
- respecting the confidentiality of information about the results of conformity assessment procedures for imported products in the same way as for domestic products so that commercial interests are protected; and
- working towards mutual recognition of conformity assessment procedures.

The TBT Agreement also encourages Members to participate, within the limits of their resources, in the work of international standards bodies and to establish an inquiry point to serve as a focal point for information on regulations, standards, and conformity assessment procedures. It also requires that Members promptly notify other signatories of proposed regulations having a potential impact on trade. Within the United States, NIST's National Center for Standards and Certification Information (NCSCI) serves as the U.S. inquiry point.

The United States government is also a signatory to other regional and bilateral trade agreements that impose similar obligations.

CHALLENGES FACING THE U.S. STANDARDS SYSTEM

No system, particularly one as complex and diverse as the U.S. voluntary standards system, is without problems. Some of the many challenges facing the U.S. system (and frequently standards systems outside the United States) include:

- **Legal challenges.**[63] There have been an increasing number of legal actions taken against standards developers. Such actions are particularly likely for standards developers that address health or safety related issues. Such suits, even when won, force standards developing organizations to incur substantial legal and other costs in defending themselves.[64] Since late 1996, there have been at least three decisions in which the court held that a standards developer does owe a duty of care to those impacted by the application of the developer's standards. The first such decision was Snyder v. American Association of Blood

---

[63] Much of the information in this section came from the writings and opinions of Amy A. Marasco, ANSI General Counsel, in particular from her article "Standards Development: Are You At Risk?" and her testimony before the Federal Trade Commission on December 1, 1995.

[64] Amy Marasco.

Banks (hereafter 'Snyder'). In Snyder, the plaintiff brought claims of strict liability, breach of warranty, negligence, and consumer fraud against the American Association of Blood Banks (AABB) alleging that he had contracted AIDS from a transfusion of blood received during open-heart surgery. A jury found AABB negligent for not recommending that its member blood banks conduct surrogate testing of donated blood for HIV. In 1998, the jury in the Superior Court of the State of Washington for the County of Benton awarded the plaintiff in Meneely v. S.R. Smith, Inc. et al. $11 million in damages, 60% of which was to be paid by the National Spa and Pool Institute (NSPI). NSPI was a standards developer that published standards for swimming pools and spas. The Court concluded that if it is foreseeable that someone could be injured in a pool conforming to a NSPI standard, then there is a duty and enough of a nexus to justify finding that NSPI's conduct caused the injury even though, as noted by the Court: the standard was voluntary; the excavation contractor did not rely on the NSPI standard; the pool did not meet NSPI standard for a Type II pool; the diving board installer did not measure the pool; and the plaintiff performed a suicide dive with his arms at his sides and not over his head.

- **Copyright issues.** Such issues are covered under the Copyright Act (17 U.S.C.). An owner of a copyright to a document, such as a standard, is generally assumed to have exclusive rights to the copyrighted work. However, in a recent decision in Veeck v. Southern Building Code Congress International Inc. (SBCCI), 241 F.3d 398 (5th Cir., February 2, 2001), the Court addressed the issue of whether a private sector standard loses its copyright protection when the standard is adopted into law or regulation. The case involved a standards developer, SBCCI, which develops and promulgates model building codes that are frequently mandated through legislative action by state and local governments. Veeck operated a nonprofit website providing information that included the texts of local building codes. Veeck purchased copies of the SBCCI codes and then published the codes on his website as local building code regulations (not as the SBCCI Code). Veeck's case revolved around the assertion that the codes became public domain when they were enacted into law. Veeck also asserted that the public is entitled to have free access to the law is so that they can comply with it. Veeck argued that statutes and judicial opinions cannot typically be copyrighted because they are considered public goods and are excluded from protection under the Copyright Act. Veeck won the case on appeal. The Fifth Circuit's decision held that, to the extent a privately copyrighted standard or code is referenced into law (particularly if it thereby becomes "the law"), the developer cannot enforce its copyright against any free (or conceivably other) distribution of it as "the law." The case was appealed, but the Supreme Court declined to hear it. Since many standards developers depend on income from the sale of standards to finance their standards development efforts, such case are of considerable concern.

- **Antitrust considerations**. The Sherman Act prohibits contracts, combinations or conspiracies in restraint of trade, and the Federal Trade Commission Act prohibits unfair methods of competition and deceptive practices that affect commerce. Standards that result from collaboration by competitors or that unfairly discriminate against certain types of products or manufacturers can be and have been subject to antitrust actions by the government.

- **Forum shopping.** This can occur because there are many standards developing organizations at the national and international level. Few (if any) in the United States can afford to participate in all potentially relevant standards developing organizations; and those with opposing views from the United States can pick standards development fora where U.S. participation is lacking. Recent examples of this have occurred in the food safety area, where ISO began work on standards that were similar to efforts already underway development within Codex, an intergovernmental standards-setting body established by the Food and Agriculture Organization (FAO) of the United Nations (UN) and the World Health Organization (WHO). Both the U.S. government and industry had been focusing their standards participation efforts and resources in the work of Codex, because of the preferential standing of Codex under the WTO Agreement on the Application of Sanitary and Phytosanitary Measures when they became aware of competing work being undertaken within ISO.

- **"Open Access" debate.** This debate, associated with government-funded research, affects documents (such as standards) that report the results of such government-funded research. Under Section 105 of the Copyright Act, copyright protection "is not available for any work of the United States Government." Such works are generally considered to be in the public domain with only a few exceptions. The open access argument generally focuses on the idea the taxpayer should not pay twice for the same research. In addition, some feel that since the works of tax-paid government employees fall into the public domain (Section 105 of the Copyright Act), the works of tax-paid government contractors and grantees should also fall into the public domain. Such a policy can have a significant effect when the funded effort is a standards development effort that results in publication of a standard. Revenue from the sales of a standard can be significant. SDOs are not likely to willingly provide free, publicly accessible copies of standards whose development has been funded in part by a government agency. SDOs, faced with the loss of revenue from the sale of a standard, could be reluctant to accept government funding and to participate in the development of a standard needed to implement a government objective. To date, the impact of such issues has not really been felt -- partly because government funding for standards development activities has been limited. Only a few agencies have partially funded private sector standardization efforts; and, given the fiscal realities currently facing the U.S. government, an SDO funding model that largely relies on significant, consistent and dependable government funding for standards activities is unlikely to be a realistic possibility in the near future.

- **Growth of consortia.** The growth in number and scope of consortia has presented a serious challenge to more traditional standards developing organizations and systems. While both consortia and traditional standards developing organizations have their place in the U.S. and international standard systems, the strengths and particularly the limitations of consortia-developed standards need to be better understood by users of such standards. Standards consortia also need to be better integrated into U.S. and international standards systems, which may require that the traditional standards developers rethink the way they do business.

- **Incorporation of patented technology or copyrighted material.** The incorporation of patented technology or copyrighted material into a standard without certain safeguards can produce an unacceptable, anti-competitive marketplace effect. Standards developers need policies that ensure that a standard does not include patented technology or copyrighted material if it gives the patent/copyright holder or any organization, country, or geographic region an unfair marketplace advantage. While it may occasionally be desirable to incorporate patented technology/copyrighted material into a standard, there is a need to balance the rights of the patent/copyright holder with the right of all interested parties to be able to implement the standard. If patented technology or copyrighted material is incorporated into a standard without the patent/copyright holder's agreement to share the patents or copyrights, then the patent/copyright holder may be the only entity able to comply with the standard. This can give the patent/copyright holder a significant marketplace advantage over any competitors. There are sometimes built-in incentives to discourage a patent/copyright holder from keeping silent about patented technology or copyrighted material that may be included in a standard until the standard is finalized. In some cases, the standard can be withdrawn, often rendering the company's patented technology or copyright relatively useless. Competitors can avail themselves of their legal rights in court. Patent or copyright holders can be forced to forgo licensing royalties under a patent misuse scenario. However, there is no easy answer for cases where a participant's failure to disclose a patent or copyright held by his organization is inadvertent. Some participants may not be aware of all the patents/copyrights held by his/her organization. This is particularly true for some organizations that hold an extensive number of different patents/copyrights.

While many organizations are working hard to addresses these challenges, they continue to pose potential threats to the continued viability and effectiveness of the U.S. standards system.

U.S. STANDARDS STRATEGY (USSS)[65]

The *United States Standards Strategy*, published in December 2005, is a revision of the 2000 *National Standards Strategy for the United States* (USNSS). The name change recognizes the globalization of the marketplace and the need for standards designed to meet stakeholder needs irrespective of national borders. The new name also reflects a standardization environment that incorporates new types of standards development activities, more flexible approaches and new structures.

The Strategy was developed by a large, diverse group of constituents representing stakeholders in government, industry, standards developing organizations, consortia, consumer groups, and academia. It acknowledges that the: "United States is a market-driven, highly diversified society, and its standards system encompasses and reflects this framework."

The USSS recognizes:
- the strength of the United States' industry sector approach to standardization that allows interested parties in a specific industry sector or cross-sector to develop standards that address problems faced by that industry; and
- the importance of government and consumer participation in the development and use of

---

[65] The full text of the USSS is available at: http://publicaa.ansi.org/sites/apdl/Documents/Standards%20Activities/NSSC/USSS-2005%20-%20FINAL.pdf

voluntary consensus standards.

The Strategy also states that:
- the development of voluntary consensus standards should be based on a preponderance of objective evidence;
- the United States needs to actively promote the consistent, worldwide application of internationally recognized states development principles, such as those listed above;
- governments should adopt compatible approaches to using standards to meet regulatory needs and partner with all stakeholders to develop standards that have global acceptance;
- all must work to strengthen domestic and international education and outreach programs that promote understanding of how voluntary, consensus-based, market-driven sectoral standards can benefit businesses, consumers and society as a whole;
- there is no one-size fits all approach to funding standards development, though the system should be supported by all who benefit from its output; and
- there is a special need to address standards in support of emerging national priorities and related issues.

SUMMARY

"Shaped over more than a century by the changing face of this nation's history, culture and values, the U.S. standards system reflects a market-driven and highly diversified society. It is a decentralized system that is naturally partitioned into industrial sectors and supported by independent, private sector standards developing organizations (SDOs). It is a demand-driven system in which standards are developed in response to specific concerns and needs expressed by industry, government, and consumers. And it is a voluntary system in which both standards development and implementation are driven by stakeholder needs."[66]

However, the U.S. standards system is also facing a number of internal and external challenges, some of which have been described in this report. While ANSI, the standards development community, industry and government are undertaking initiatives to address some of these challenges, those who use and benefit from standards need to understand that standards are essential to preserving the quality of life in the United States and competing effectively in global commerce. Effective U.S. participation in and support of both the U.S. and international standards arenas has become and will remain a prerequisite for stimulating U.S. economic growth, safeguarding health and safety, protecting the environment, and ensuring U.S. national security.

---

[66] American National Standards Institute, "Overview of the U.S. Standardization System: Understanding the U.S. Voluntary Consensus Standardization and Conformity Assessment Infrastructure," Washington, DC 20036, July 2005.

APPENDIX I – ACRONYM INDEX

NUMERALS
3GPP2 — Third Generation Partnership Project 2
3G — Third generation

A
AABB — American Association of Blood Banks
AACC — American Association of Cereal Chemists
AAR — Association of American Railroads
AASHTO — American Association of State Highway Officials
ABNT — Associação Brasileira de Normas Técnicas (Brazil)
ACCSQ — ASEAN Consultative Committee on Standards and Quality
ACGIH — American Conference of Government Industrial Hygienists
AENOR — Asociación Española de Normalización y Certificación (Spain)
AFNOR — Association française de normalisation (France)
AIA — Aerospace Industries Association
AIDMO — Arab Industrial Development and Mining Organization
AMN — MERCOSUR Standardization Association
ANS — American National Standards
ANSI — American National Standards Institute
AOAC — Association of Official Analytical Chemists
AOCS — American Oil Chemists Society
API — American Petroleum Institute
AREA — American Railway Engineers Association
APEC — Asia Pacific Economic Cooperation
ARSO — African Regional Organization for Standardization
ASEAN — Association of Southeast Asian Nations
ASME — American Society of Mechanical Engineers
ASTM International — Formerly the American Society of Testing and Materials
ATCR — ASEAN Common Technical Requirements
AUTOSAR — Automotive Open System Architecture

B
BNSI — Barbados National Standards Institution
BSJ — Bureau of Standards Jamaica

C
CableLabs — Cable Television Laboratories, Inc.
CAC — Codex Alimentarius Commission
CAN — Andean Community
CARICOM — Caribbean Common Market
CCMSC — Caribbean Common Market Standards Council (now CROSQ)
CEN — European Committee for Standardization

CENELEC	European Committee for Electrotechnical Standardization
CIS	Commonwealth of Independent States
CODEX	Codex Alimentarius Commission
COGUANOR	Comisión Guatemalteca de Normas (Guatemala)
COHCIT	Consejo Hondureño de Ciencia y Tecnología (Honduras)
CONACYT	Consejo Nacional de Ciencia y Tecnología (El Salvador)
COPANIT	Comisión Panameña de Normas Industriales y Técnicas (Panama)
COPANT	Pan American Standards Commission
CROSQ	CARICOM Regional Organization for Standards and Quality
CTFA	Cosmetic, Toiletry and Fragrance Association, now known as the Personal Care Products Council

D/E

DGN	Dirección General de Normas (Mexico)
DIGENOR	Dirección General de Normas y Sistemas de Calidad (Dominican Republic)
DOD	Department of Defense
EASC	Euro Asian Council for Standardization, Metrology and Certification
EFTA	European Free Trade Association
EIA	Electronic Industries Association
ESO	European Standards Organization
ETSI	European Telecommunications Standards Institute
EU	European Union

F

FAO	Food and Agriculture Organization
FCC	Federal Communications Commission
FDA	Food and Drug Administration, Department of Health and Human Services
FHWA	Federal Highway Administration
FONDONORMA	Fondo para la Normalización y Certificación de la Calidad (Venezuela)
FTC	Federal Trade Commission

G/H

GDBS	Grenada Bureau of Standards
GNBS	Guyana National Bureau of Standards
HS	Harmonized System

I

IAAC	Interamerican Accreditation Cooperation
IANOR	Institut algérien de normalisation (Algeria)
IBNORCA	Instituto Boliviano de Normalización y Calidad (Bolivia)
ICC	International Code Council
ICONTEC	Instituto Colombiano de Normas Técnicas y Certificación (Colombia)
ICT	Information and communication technologies
IEC	International Electrotechnical Commission
IEEE	Institute of Electrical and Electronics Engineers
IEEE-ISTO	IEEE's Industry Standards and Technology Organization

IETF	Internet Engineering Task Force
INDECOPI	Instituto Nacional de Defensa de la Competencia y de la Protección de la Propiedad Intelectual (Peru)
INEN	Instituto Ecuatoriano de Normalización (Ecuador)
INN	Instituto Nacional de Normalización (Chile)
INTECO	Instituto de Normas Técnicas de Costa Rica (Costa Rica)
INTN	Instituto Nacional de Tecnología, Normalización y Metrología (Paraguay)
IPQ	Instituto Português da Qualidade (Portugal)
IRAM	Instituto Argentino de Normalización y Certificación (Argentina)
ISO	International Organization for Standardization
ITU	International Telecommunications Union

J/K/L/M

MERCOSUR	Mercado Común del Sur
MIFIC	Dirección de Tecnología, Normalización y Metrología (Nicaragua)
MILSPEC	Military specification
MOU	Memorandum of Understanding
MRA	Mutual Recognition Arrangement
MUTCD	Manual on Uniform Traffic Control Devices

N

NATO	North Atlantic Treaty Organization
NC	Oficina Nacional de Normalización (Cuba)
NBS	National Bureau of Standards (now NIST)
NEMA	National Electrical Manufacturers Association
NFPA	National Fire Protection Association
NIST	National Institute of Standards and Technology
NSF	NSF International, formerly the National Sanitation Foundation
NSPI	National Spa and Pool Institute
NTTAA	National Technology Transfer and Advancement Act

O/P

OECD	Organization for Economic Cooperation and Development
OMB	Office of Management and Budget

G/R/S

RRT	Radio transmission technology
SABS	South African Bureau of Standards
SADC	Southern African Development Community
SADCSTAN	SADC Cooperation in Standardization
SAE	Society of Automotive Engineers
SBCCI	Southern Building Code Congress International Inc
SCC	Standards Council of Canada
SDO	Standards Developing Organization
SLBS	Saint Lucia Bureau of Standards

STANAG   NATO standards agreement

T
TAG   Technical Advisory Group
TAPPI   Technical Association of the Pulp and Paper Industry
TBT   Technical Barriers to Trade
TIA   Telecommunications Industry Association
TMC   USNC Technical Management Committee
TTBS   Trinidad and Tobago Bureau of Standards

U/V
UL   Underwriters Laboratories
UN   United Nations
UN   UN Economic Commission for Europe
UNI   Ente Nazionale Italiano di Unificazione (Italy)
UNIT   Instituto Uruguayo de Normas Técnicas (Uruguay)
UPU   Universal Postal Union
USNC   U.S. National Committee to the IEC
USNSS   National Standards Strategy for the United States (now the USSS)
USSS   U.S. Standards Strategy

W/X/Y/Z
W3C   World Wide Web Consortium
WCO   World Customs Organization
WHO   World Health Organization
WMO   World Meteorological Organization
WTO   World Trade Organization